FEET, FLIPPERS, HOOVES, AND HANDS

BY MARK J. RAUZON

For Karla Kral

Acknowledgments

Special thanks to the following people for the use of their photographs: Craig S. Harrison for the gerenuk on page 3, the baby impala on page 5, and the giraffes on page 17 — photographs © Craig S. Harrison; Warren Garst/ Tom Stack and Associates, Inc., for the leopard and bush pig on pages 6 and 7 — photograph © Warren Garst.

LOTHROP, LEE & SHEPARD BOOKS   **NEW YORK**

BASILISK LIZARD ▲　　　　　　　　**▼ ORCAS (KILLER WHALES)**

GERENUK

MEERKAT

Almost all animals need to move around to find food, to escape from danger, and to find mates. But exactly how an animal gets around depends on where it lives, what it eats, and who its enemies are. Feet, flippers, hooves, and hands are designed to help animals survive.

◄ **BROWN PELICAN**   ▲   **BABY IMPALA**

FEET

There are many different kinds of feet—clawed feet, webbed feet, feet with hard hooves, soft padded feet. Each kind of foot has its own use. Clawed feet make a squirrel an expert climber. Hoofed feet enable antelopes to run fast over hard ground. Webbed feet help pelicans to swim, and a tiger's padded feet, or paws, allow it to creep silently. But a foot's most important job is to help an animal balance on its legs. Legs and feet work together to help animals stand up, run, climb, or jump.

LEOPARD CHASING BUSH PIG

Back legs and feet work differently from front legs and feet.
Like most four-legged mammals', the leopard's strong hind legs

and feet provide the power for chasing prey. Since front paws are designed for grabbing and holding prey, they must be bigger than the rear paws and have sharper claws.

CALIFORNIA MOLE

Moles use their strong, clawed front paws like shovels when digging for worms and insects. With their back feet, they push the dirt out of the way and pack it down.

Like giant rabbits, kangaroos have long back feet for bounding over the ground. A big "roo" can jump over thirty feet in a single bound. But its front paws are short and flexible to gather and hold food.

Eagles, hawks, and owls have clawed feet for catching animals. They dive from great heights, with their feet outstretched, and surprise their prey. This bald eagle has needle-sharp claws called talons on its scaly feet to grip slippery fish and birds.

BALD EAGLE WITH RUDDY DUCK

◀ **RED-FOOTED BOOBY**   ▲ **GIANT TREE FROG**

Some animals need feet with claws in order to climb and cling to tree trunks and branches. Birds such as boobies grip branches with their clawed feet. A fruit bat hangs upside down by hooked claws on its back feet and uses hooks on its fingertips for "walking" under branches. Even insects have clawed feet for climbing.

Not all animals need claws for climbing. A tree frog has sticky fingers and toes to hang on to leaves.

FLIPPERS

Flippers are special feet adapted for swimming. Sea turtles have powerful front flippers to propel them forward and rear flippers to steer them as they swim underwater. Female sea turtles also use their huge front flippers to drag themselves out of the ocean and onto the dry beach and their rear flippers to dig nests in the sand to hold their eggs.

HAWKSBILL SEA TURTLE

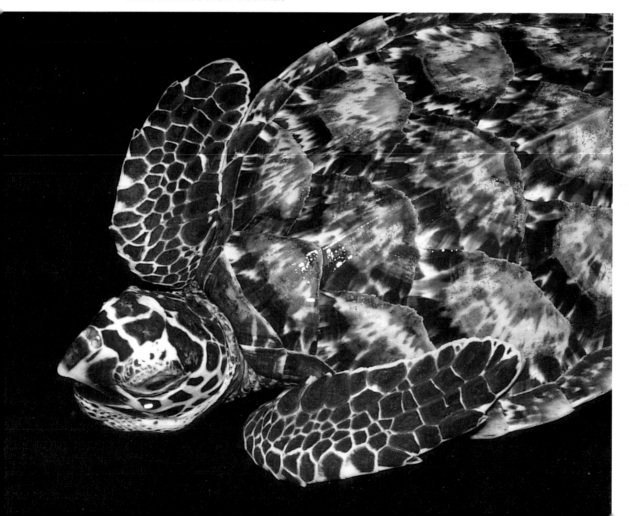

ELEPHANT SEAL

Marine mammals such as sea lions and seals have flexible flippers too. They use their front flippers to pull themselves out of the water as well as to swim. Their flippers are also useful for flipping cool sand over their backs or scratching an itch.

BLUE WHALE

Whales, dolphins, dugongs, and manatees have front flippers, but in the back they have flukes. They once had rear flippers, but over time the back flippers disappeared because the animals could swim faster with flat flukes on their tails.

The front flippers steer the course while the powerful flukes push the water from behind. Flippers and flukes are also used to communicate. Whales and seals slap the water with their flippers and flukes to make their presence known to one another.

▲ ELK

GIRAFFES ▶

HOOVES

Hooves are very strong toenails, so animals with hooves are always on their toes. Hooves evolved because animals can run faster for long periods on tiptoe than on flat feet. Some kinds of antelope can run forty miles per hour to escape an enemy. When speed is not enough, a swift kick from a sharp hoof is also good protection.

The hoof is like a shoe around the foot. The inside of the hoof is cushioned with a thick pad of hard skin. Hooves are made from keratin, the same material as horns and hair. Like horns and hair, hooves are always growing and always wearing away.

ZEBRA

Hooves can have one or two toes. One-toed animals such as zebras and horses have round, heavy hooves. They can run fast for long distances, galloping away with each foot barely touching the ground.

Split-hoofed animals like goats and sheep run slower, but they can climb, dig up roots, and scratch an itch better because their feet are more flexible. Most split-hoofed animals also have horns, antlers, or tusks to help protect themselves.

DALL SHEEP

HANDS

Some animals have flexible front feet especially designed to help them push, pull, grab, grip, pick up, feel, and throw things. These special feet are called hands. Hands have four or five fingers that may be tipped with protective claws or nails. Raccoons and sea otters grab food with their hands and hold it while eating.

SEA OTTER

HUMAN HOLDING JUMPING MOUSE

Some hands have a special first finger called a thumb. Animals with thumbs, such as chimpanzees and humans, can do things with one hand that thumbless animals need both hands to do. Try to pick up this book without using your thumbs.

Hands with thumbs evolved for life in the trees. Animals such as monkeys and apes, koalas, opossums, and lemurs can climb better because of their thumbs. They can also use their thumbs to pick fruit and hold their babies.

◀ **WHITE-FACED CAPUCHIN MONKEY**　　　　　▲ **FRUIT BAT**

The human hand contains over twenty-five small bones that work very well together. Hands and feet account for more than half the bones in the human body. Other animals have fewer bones in their hands and feet, so they are not as flexible. Instead, they have long necks or tails and strong jaws to help them get food and hold their young.

Wings are arms and hands especially modified for flying. Bat wings are very large hands. Leathery skin stretched over the bat's long slender fingers enables it to catch insects on the wing.

FOX SQUIRREL

CALIFORNIA SEA LION

EUROPEAN BOAR

LOWLAND GORILLA

No matter how they move — by running, jumping, swimming, swinging, climbing, or flying — animals all over the world can find food and escape from their enemies because of their feet, flippers, hooves, and hands.